Principles for a Successful Tech Start-up Company with Little or No Money

Copyright ©2023 by Ronald S. Johnson

All rights reserved. No part of this book may be reproduced by any mechanical, photographic, or electronic process, or in the form of a phonographic recording; nor may it be stored in a retrieval system, transmitted or otherwise copied for public or private use - other than for "fair use" as brief quotations embodied in articles and reviews – without written permission of the author.

ISBN: 979-8-218-26497-0
Published and printed in the United States of America
First Edition Printed 8-13-2023
Contributing Editor: Aziz Makhani

Contents

Acknowledgements 8
PREFACE 9
SO, YOU HAVE NO START-UP CAPITAL 11
   Money happens when you do the right thing 11
   Bootstrapping 11
   Running a company is all about resource management 12
   How can I expect others to help when I cannot pay them?.. 12
   People want to help you succeed 13
   One plus one equals three 14
   Leverage 15
   The benefit of being a student 16
   Focus on looking adorable 17
   Being well-balanced 18
   " But I want to run my company in stealth mode" 18
STARTUP PRINCIPLES 19
   Forget about the money 19
   Always be moving forward 20
   There is always another way 20
   There is no failure 20
   Focus on revenue 21
   Friends do business with friends 21
   Don't go formal until you need to 22
   Technology is like lettuce 23
   Going into the mouth of the lion 23
   Learn to cooperate and "get along" 24
   Greed kills 25
   "I am afraid someone will steal my idea" 25
   Start where you are 26
A WORD ABOUT INVESTMENT CAPITAL 27
   Raising capital is not an event – It is a way of thinking 28
   You need to be invited to the dance in order to play 28
   Honesty is the best policy 29
   A word about giving early equity to participants 30

Should I use debt financing? 32
ELEMENTS OF RISK 34
   Why do we care about risk? 34
   Old technology risk 35
   Incomplete information risk 35
   Channel risk 35
   Technology leapfrog risk 36
   Pricing risk 36
   Disappearing margin risk 37
   Alliance risk 37
   Team communication risk 37
   Speed/Motivation risk 38
   Divergent/Poor strategy risk 38
   Take the money and run risk 39
   Take the technology and run risk 39
   Next stage of capital risk 40
   Dilution risk 40
   Liquidity/Exit risk 40
   Product liability risk 41
   Reverse engineering risk 41
   Improper disclosure risk 41
   Employee risk 42
   Arrogance risk 43
   Improper balance risk 43
NEEDS LIST TO HELP YOU GET STARTED 44
   Instructions: 45
   Elevator pitch 48
   Company exclusive bank account 49
   Company exclusive credit or debit card 49
   Business plan 50
   Strategic plan 51
   Operating/Financial plan 51
   Product/Generation expansion plan 53
   Marketing plan 53

Marketing partner 54
Development partner 54
Purchasing high-cost equipment 54
Raw material 55
Office/Lab/Production space 55
Test equipment 56
Referrals 56
Software/Hardware engineering 56
Web help 56
Prototype 57
Accounting system/Accountant 57
Partner 57
Advice 57
Advisory Board member 58
Sales Presentation Material 58
Intellectual property 58
A DISCUSSION ON CORPORATE VALUE 59
   You need a corporate plan 59
   Stock value and the achievement of milestones 60
CONCLUSION 62
Appendix A 63
   Sample Company Concept Paper (fictitious) 63
Appendix B 67
   Equity Growth Chart 67
Appendix C 68
   Operating/Financial Plan Overview 68
Appendix D 69
   Operating/Financial Plan Resources 69
   Short Operating Plan 69
   Full Operating Plan 69

Acknowledgements

I would like to thank Jerry Joyce of the Interim Controller for his mentorship over my years of working in Silicon Valley. Without his friendship and mentorship, this book could not have been written.

I would also like to thank Aziz Makhani for his encouragement and insights into entrepreneurship and specific guidance for young companies. Also, for his help in editing this book.

Peter Allison will always be remembered as the consummate legal counsel, fund raiser and strategic player, who applied his trade so effectively in the real world of mentoring and financing the growth of young companies.

PREFACE

In the course of my work, I have learned that there are methods and principles that can accelerate a company's growth and methods and principles that can slow and potentially stop a company's growth. It is the objective of this book to ignite the fire in your company that will propel it forward at a rate faster than you are currently experiencing.

The essential element that will make your company fail or succeed is within you. It is a simple matter of perspective. Sounds easy, right? So, my objective here is to perhaps give you a different perspective on the way you think about your company, your competitors, your customers and your investors. Right thinking is the basis for right performance.

Although this is predominantly a book about starting your tech company with no money, you will likely need additional capital to help fuel your growth at a later stage. So, once we get past the "no money" principles, we take a look at what early-stage investors look for in a start-up company. We will look at how to evaluate your company from the perspective of an outside funding source and how to eliminate the fatal flaws often revealed in your investor presentations.

I have spent the past 48 years working with hundreds of entrepreneurs and start-up companies. This has included full time and consulting positions as a CFO, Controller, Director of Finance, Corporate Planner and Coach.

I also co-founded a seed-funding investment banking firm in Spokane, Washington. And it was through this experience that I, for the first time, had to look at our young company applicants as an investment and fully understand the ecosystem of a venture capital company, where it gets its money, how it makes money, how the young company can succeed in the capital arena and how they can fail miserably. This was truly an eye-opening experience which would later provide the basis for this book.

This book is a little less of a topical book and a little more of a conglomeration of my experiences. You could probably place this book in the genre of "things I wish I had known earlier in life". It could be that you will only take away one or two concepts that will change your performance for the better. But they may be key elements to your success. And for such a result, I would be grateful, and satisfied that the book had reached its objective.

You will not necessarily find solutions to all of your problems in this book. It is your task to provide the solutions. My hope is that the contents will open your eyes to another perspective that is important for you to have, to become successful in your venture.

SO, YOU HAVE NO START-UP CAPITAL

Starting a company with no money is a great place to be. You may see this as a difficult position in which to be. But to me it was always the consummate challenge; how to get to the top with no immediate resources. Your biggest assets here are your ability to think creatively, and then to act on those creative thoughts.

Money happens when you do the right thing. Getting money, whether revenue or investment capital, is all about doing the right thing. If your company is strategically focused, as well as correctly balanced and aligned, investment capital will want to come to you as they see that there may be an opportunity to make more money with their investment. And your products or services will be exactly what your customers are looking for. Money is not the action; it is the result of right action. The objective here is to make your company so attractive, that the money comes to you, from investors and customers.

Bootstrapping - Yes, many companies started up successfully before venture capital companies even existed. So how did they do it? It is called bootstrapping. That is taking one step at a time and recognizing the benefits of that one step to take you to the next step. You can visualize it as climbing a ladder one step at a time.

Bootstrapping describes a situation in which an entrepreneur starts a company with little or no capital, relying on their prowess and personal money other than outside investments. This is a slower start-up method. With the investment money, you could move faster. But investment money can be costly at this stage. Get used to accomplishing your goals without outside money and you will be doing yourself, your company and your future investors, a great service.

But of more importance here, is that you are learning not to solve problems by throwing money at them. I have seen this many times over in my career. A young company gets venture capital and they immediately start to solve their problems by throwing money at them. This leads to a waste of resources and often leaves them without the needed capital to accomplish their future milestones. To be fair, I have also seen the startup that is afraid to spend money and slows their progress toward their milestones as well. Learning how to spend money when it is needed and when the spend will provide the leverage you need to accomplish your milestones, is a key to success.

Running a company is all about resource management – Ask any young entrepreneur about what they really need right now to move the company forward, and 90% of the time, you will hear "money". Seems reasonable. But having this perspective will focus you in the wrong direction from the start. You don't really need money. What you really need are the resources that the money can buy. So, let's look at how to acquire the resources.

How can I expect others to help when I cannot pay them? – We all have different motivations for where we spend our free time. Some folks require compensation for every effort they make on another's behalf. Some others are more congenial and benevolent and may follow their interests with their free time. If they are interested in what you are creating and want to become a part of your effort, and if they see a way they can help to meet your needs, they may spend some time and not expect compensation. This is not about taking advantage of others. If you are receiving income from their efforts, you would want to share the wealth. But initially, they will likely spend as little amount of time as they can spare without expecting monetary compensation.

It is up to you to show them, and the world, the potential benefits of your work and share your vision with them. If you invite them in so to speak, and show them how they can fit into your effort with their minimal effort, you have accomplished the first rung of the bootstrapping ladder. You will see the speed of your effort pick up tremendously. Be grateful for them and reciprocate with your helping them or giving back in some way, when an opportunity arises.

> *Real-Life Experience – An entrepreneur came to me asking my advice on ways to proceed forward with his company. When we met at the door, he presented me with a bottle of wine as an advance thank you for my time and advice. I was a little taken aback as no one had ever reciprocated for my offers of free advice before. We talked for an hour or so and became great friends through this discussion. I gladly provided many more hours of free advice because I knew he was a grateful soul and had a good perspective on the value others provided him.*

People want to help you succeed – Most people do want to see you succeed as an entrepreneur. And if they can help without taking a lot of their time, they are most happy to assist. So, the key is to ask for help but only where you honestly need the help and where you feel you will not be imposing on them. I found it is always helpful of you can return a favor for a favor. If someone agrees to help you for free, return the favor with a handwritten note of thanks, a bottle of wine, concert tickets, a good cigar, maybe mowing their lawn or maybe you can help them with another need they have. So, look at your interactions with others as to how you can help them in return.

But the key here is to only ask for a little. Do not impose on them and expect them to have your passion for your company.

> ***Real Life Experience*** *– A friend of mine was starting a business and asked me for help putting together a financial forecast. I saw the great success he was having even though he had essentially no budget and had not taken in any outside capital. He enrolled friends who had specific skills he needed by asking for only a little. I was only too happy to help him construct a financial forecast. I was so impressed with the progress he had made without funding that I wanted to be a part of his success. I would typically charge $5,000 to $6,000 for this effort, but was happy to help him out at no cost. All he had to do was ask. And remember, friends do business with friends.*

One plus one equals three
Building your team, and motivating this team, is likely the most important task you have as an entrepreneur. Your team is the company, and your company's character will evolve as you bring in talent to help achieve your goals.

We are all cut from a different cloth. This enables each one of us to have unique abilities and perspectives. These abilities and perspectives are not only desirable in a start-up, they are a necessity. When you bring additional contributors into your company, they can expand your horizons greatly and can influence your success. Be open to other perspectives with the people who you enroll in your project. Divorce the company from you. It is a stand-alone effort that requires the input and talents of many. Focus on doing what is right for the company and not what you feel is compatible to your point of view.

Real Life Experience - *One startup that I counseled had a hardware/software engineer who had developed a bench prototype to show the ability of his circuitry and software to perform the task at hand. He thought he was ready for funding.*

I suggested building the hardware into a physical prototype so the investors could get a better feel for the eventual product. This would also provide pictures for the marketing material we would need to prepare for the market rollout. He had never thought of this need as it was down the road and not in his area of expertise. I had been down that road and was able to take the prototype to the next level for uses he hadn't considered. My input carried him way down the road in terms of corporate development.

Real-Life Experience - *Another simple example of one-plus-one equals three occurred one day when a fellow entrepreneur stopped by my house. I mentioned that we had a problem with rain getting in the South facing door and soaking the interior floor. My potential solutions included building a mud room and reorienting the door or perhaps pulling out the door and frame altogether and replacing it with one that sealed better. He looked at the door and casually said, why don't you just set in a storm door that seals well. His solution was so simple, and by the way, it has worked well. I would never have arrived at the optimum solution myself. His input was invaluable. One plus one equals three. If you build a good team that works well together, you will see this optimization occurring many times over.*

Leverage - To successfully implement leverage in your company, you need to recognize and promote the concept of leverage in every area of your business. Leverage is simply the concept of spending a buck, and getting two back. Every dollar spent has to be applied to its greatest benefit.

One way to realize this leverage is to have others around you who can provide a benefit you cannot provide, and then managing those assistants to the defined end. So, you may need to practice enrolling and motivating others to work with you to whatever level of effort they can contribute.

This will mean getting out and meeting and talking with others. This is often difficult for entrepreneurs who may tend to be introverted. So, your first step may be to enroll someone to help who is an extrovert. They have a skill or contacts you may not have. They will be invaluable in helping you meet your objectives along the way. Work with them to decide what message you want to send to achieve your next step up the bootstrapping ladder. What is your immediate next need? Who can you approach to help you meet this need? What will you need to help communicate this need and potentially fill it?

The benefit of being a student – The university setting offers a plethora of resources. Not only in terms of lab space, mechanical devices, laboratories, etc., but also in personnel resources as well. If you are a software engineering student, you may want to find a marketing student to fulfill a "Director of Marketing" position. Look to your school of business for a student to be your "Director of Marketing". No, they will not likely be the most experienced or qualified to represent the company, but their different perspective will be an invaluable aid in getting your product or service to market.

Start your company when you are in school. There is nothing better on a resume than to show that you ran a company while in college. The clue here is to only ask for a little from each of your fellow students. If you ask for all of their spare time, they will expect to be reimbursed for it. But an occasional meeting or research project is a very reasonable request.

As a venture capitalist, I would never invest in a person or product. But I would always invest in a team and a company.

Focus on looking adorable – Our Western approach to raising capital tends to be very masculine by its nature. That is, we tend to want to slay the dragon, go out and get the money. Get the road show together and contact the investors. But there should also be a feminine side to your capital raise. That is looking adorable and attracting investors to you. If investors can see an opportunity to successfully invest in your company, they will be attracted to the investment opportunity. So, consider including a feminine element to your capital raise strategy.

For example, have a contributor write four articles about your technology, and indirectly of course, about your company. Then submit them to local and regional newspapers, and perhaps relevant websites, one at a time over the course of several weeks or months. Show your successes and paint the bright future. Investors who read those media may take an interest and reach out to your company for a possible investment opportunity.

Stakeholders in your company want to see a functioning, operating company. This is most easily achieved by your building a company that can show results in moving forward even without money. You will need to come up with strategies that draw the money to you. Strive for excellence. Be the best in your industry and be able to express why you are the best.

Being well-balanced - Another element in "looking adorable" has to do with balance. As I have mentioned earlier, we are all cut from a certain cloth. Engineering and scientific people tend to look at the world through the glasses of their experience, while an accountant or marketing person may have a completely different view of the world. And this view is also very important in developing a balanced plan going forward. Remember that you may be predisposed to focus on only one or two areas of your business, neglecting others in which you may not have experience. This is where outside help, or the help of your partners, can make a significant improvement in the strength of your presentation.

Focus on looking adorable to your investors and the money will come to you.

" But I want to run my company in stealth mode" – Many companies die from lack of exposure to their market. Very few have ever succeeded with a stealth mentality. To successfully bootstrap your company, you will need to become the evangelist. But how then do you fight off the early competition? Become better, more agile and look ahead and plan for the moves of your potential competitor. Don't try to avoid competition. Plan for it and face it head-on. Go faster and smarter and you will win in the end.

STARTUP PRINCIPLES

I would consider the following principles to be key factors in starting up your company, with or without funding. I will highlight the principle and give a short explanation as to why it is important.

Forget about the money - Your company needs to move forward every day regardless of the resources currently available to it. Forget about the money. Investment capital is most often not available, and to sit around and wait for it will be your demise. This book will help you put together the perspective and plan to get your company moving forward without outside capital.

> *Real-Life Experience* - *I was tasked with taking entrance interviews for our venture capital company. I interviewed many entrepreneurial teams and learned that there are certain fatal flaws that often were revealed through these interviews. The most in-your-face failure was when an entrepreneur finished their presentation, and would sit back and say, "now all I need is the money". At that point, I would likely wrap up the interview and dismiss the company from the running. Sounds kind of harsh you say? Maybe.*
>
> *But I could just see that company president coming back to me when the company had run out of money and they had not met their objectives and citing that phrase, "all I need is the money to keep going". Now I would have to make the decision will I put good money after bad? I never wanted to get to that situation. So, I looked for entrepreneurs that were moving forward, meeting their objectives, regardless of how much money they had. They had the resourcefulness to find a way to move forward without money. So, in short, the success of their company was not dependent on me. My*

> *capital was meant to help them move forward faster and perhaps gain a greater market share than if they had to move forward organically.*

Always be moving forward – The concept here is one of what I call "tilting-in". You will have to know in your own mind that your company will always succeed, with or without the money. Money will only help you achieve your objectives faster. You will want to show your investors that you will always be moving forward regardless of the capital you have to operate.

There is always another way – Think creatively here. Is there another way to accomplish the task at hand? For instance, maybe you need a CNC laser cutting machine to build your prototypes. CNC machines can be expensive. Do you need money to buy one? Not necessarily. Find who has one and see if they will help you with a few prototypes, or maybe an initial production run or perhaps ongoing production at a later date.

Or maybe you need office space. Look for a landlord or tenant who has excess space available. Maybe an extra office in back that is not being used, additional space for storage or a small production facility that you can rent for little or no money on a month-to-month basis.

There is no failure – Every effort you make is moving you forward. Even if you make a phone call and end up in voice mail, it is a success and you need to begin to look at all of your efforts in this regard. If you approach someone with a proposal and they turn you down, it was a success. There is no failure. You are always moving forward. Keep a positive mental attitude and focus on the end result. Nothing can stop you from achieving your goals if you believe in them.

Focus on revenue – I have met several entrepreneurs who have enhanced their way to failure. They keep building their product better and better, adding more functionality, etc. But they have no concept of how to generate revenue with it. Some live on investors' money until failure, again with no concept of what it takes to generate revenue. Set a plan to get to revenue and review your progress often against this plan.

> *Real-Life Experience* - *I once counseled with a software engineer who had developed a small program which became of great value to a particular governmental sector. I asked him how much he was charging for his software and he admitted that it was introduced as freeware and these "customers" were his "friends". He hadn't thought of charging them for his software. His effort was to frantically keep developing enhancements to the software with a plan to someday make money with it. But he never perceived his product to being of high enough value to ask others to pay for it. In spite of my counsel, the last I heard is that he was still developing in his spare time with no expectation of revenue.*

Friends do business with friends – Throughout my career, I have found this principle to be true. Fully half of my clientele came to me as a result of a friend or a friendship. So, this one is easy. Focus on making friends wherever you go. Never burn a bridge.

> *Real-Life Experience* – *The President of The Interim Controller, my consulting referral agency, kept copious records of everyone he met. He made an effort to get their email, phone number and sometimes their birthday and would spend some time each day reconnecting with all of his contacts in rotation. He kept track of their interests and would always have a story to tell, or an article to forward to*

> them, to keep the communication and friendship alive. He was a good-looking very amiable Irishman who would call you on your birthday and sing happy birthday to you in his Irish brogue. Sounds a little strange you may say. Perhaps, but it worked wonders for him. He worked his business from referrals only and was never short on business opportunities. Friends do business with friends. So, make an effort to make friends and then keep in touch.

Don't go formal until you need to – Some entrepreneurs thought it was proper to incorporate, get an IRS EIN number, business license, etc. before they even started to plan their business. This is really unnecessary for most startups. My preference has always been to wait until you have a need for such identifications before setting them up.

But you can start a viable business with a personal credit or debit card that is being used exclusively for the business. And you won't need to formalize your business until it is time to file your taxes, to report your profits or to claim your early-stage losses.

If you wish to operate under a fictitious business name, that is to accept payment to the name of the business, the bank will require you to get a federal EIN number, state tax ID, local business license and a copy of your formation documents.

If you are a retail establishment, you will need to have a local sales tax ID before you sell products and before you collect sales taxes from your customers. Taxing authorities make it very clear that you are not authorized to collect sales tax unless you have first registered with them, giving you the authorization to collect and remit sales tax.

Technology is like lettuce – Once technology is discovered or patented, it has a limited useful life. Begin moving forward immediately.

> *Real Life Experience* – Our VC team would frequently visit research facilities around the country looking for technologies that needed to be rolled out to a waiting market. We ran across one very impressive technology. We did some market research on the potential product and found that there was a new technology in the marketplace which was serving the market and was in fact, a later generation of the product we were considering for funding. The researcher had worked on this specific technology for 9 years, it was his baby. But it would never get to the marketplace as he had envisioned. He sat on his discovery for too long and a later generation technology got to market first. Don't let your product or service age. Move it out to a waiting market.

Going into the mouth of the lion

As an entrepreneur, you will often be faced with issues that require you to resolve, and perhaps to go where you do not feel comfortable. Being bold is a skill that you will want to acquire as you progress forward. Be bold enough to step out of your comfort zone and confront issues that need resolution. Let me give you an example of this.

> *Real Life Experience* - My former VC firm was retained by a client to prepare them for funding and to present them for funding to a larger VC firm. We made it clear up front that they were not ready for funding and needed an operating plan and a financial forecast before we could consider moving them forward. They enrolled us to assemble this information. So, we were still in an early stage with them. They got impatient and filed a lawsuit against us because they didn't

have their immediate funding. My partner thought about the situation for a few days then one day came into my office and said, "let's go, we are going to visit the client". "What!" I said "He has filed a lawsuit against us and you want to go visit him?" "Yes, let's go".

We walked into the company and stopped at the receptionist's desk. She was obviously a little flustered to see us, but agreed to call the president of the firm and let him know we were there to see him. We met him in the conference room and yes, the situation was tense. He didn't know if we were there to be violent or what.

My partner started the dialog by saying that we were sorry for the miscommunication and didn't want to let it interfere with our relationship going forward. He pulled out a check for the amount of our initial fee and paid it back to him. He was a little shocked as we shook hands and left.

I hadn't considered it at the time, but my partner knew it wasn't worth the cost of the legal fight to go forward with the lawsuit. So best to return his deposit and call it done. This was a very wise choice. And it took the guts to confront the aggressor and settle the issue with a cool head. My reaction was to pull back, stay away and fight. His reaction was to go into the mouth of the lion and do the right thing. It took guts and was forever an example to me of doing the right thing even though it was hard and not a natural thing to do.

Learn to cooperate and "get along" – As a young company with little or no capital, you cannot afford to get embroiled in a lawsuit. Legal fees would drain all of the resources you need to move forward. And even a company that had received their first round of funding would see the investors funding drained with no progress for the company moving forward. If you want to kill your young company, go ahead and challenge the giant. You will never get an investor interested again. And your startup days will likely be over. So be careful about throwing your weight around and challenging companies and individuals who have infinitely more resources than you have. Understand your negotiating position at all times.

Personally? I believe it is good Karma to go so far as to help your competitors and look out for the interests of your suppliers. It will come back to you many times over.

Greed kills – The biggest destroyer of start-ups is greed. "I'll agree to help you but I want X% of the company". Run from these proposals. In the early stage of a start-up, the contributors have to be involved because they have a passion to see the technology roll-out and the company succeed. My opinion is that there should not be any future ownership promises made at this stage. Until an outside investor or customer takes an interest in your company, you have no value. You are worth zero dollars. You have nothing to grant or promise. The pie will get sliced up once serious capital agrees to invest and any equity promises made prior to that time will likely become irrelevant anyway. So, to our prior point, learn to get along. The idea here is let's work together to see what we can create. If it has value, we all win.

"I am afraid someone will steal my idea" – In my 48 years of experience, I have never seen a situation where someone else stole an idea from an entrepreneur. Not saying it doesn't happen. I am sure there are examples of idea stealing. Sure, Steve Jobs and the mouse idea stolen from Microsoft. Yeah, I get it. It does happen. But many ideas have failed through being hidden and more often than not, if you want to remain secretive, your product or service offering will also remain secretive. Which

25

isn't the reason you are in business. The idea here is that you have to move faster than could any competitor. You need to be better, more informed and more knowledgeable than your competitor. Move it forward quickly.

If you are successful, there will be someone who will copy your product or service and begin competing with you. Be ready for this situation to occur and have a strategy to defend your position before the situation arises.

Start where you are – Get organized now through the use of the later chapters in this book. These tools will help you evaluate where you are and where you need to go next. But remember, keep it moving!

We will look deeper into what is needed to raise capital in the next chapter of this book. But for now, we are doing it without money, right? Let's move on...

A WORD ABOUT INVESTMENT CAPITAL

Venture Capitalists have to be good at one thing, that is making money for their investors. And typically, that is their strong suit. They know how to take in large amounts of capital and make money with that capital. So, understand that this is their focus in giving you the money. They expect huge increases in value early-on in your corporate development.

If your intent is to increase the value of your stock with your startup, you will have no problem getting along with your capital source. If your objective in starting your company is to "take your baby all the way", to become famous for your company, or if your identity is wrapped up in your position with the company, don't take outside investment capital. Your objectives may eventually clash and there will be an upset.

When experienced and knowledgeable equity capital comes into your company, it will come with many conditions. One of those conditions, written or unwritten, will be your performance. The management team is responsible for putting together a budget and revenue forecast, and then guiding the performance of the company to make that forecast happen. Should management consistently fall short of the objectives established, they are in danger of being replaced with another management team who the board of directors believe can make the forecast happen.

So, if you are the president/CEO of the company, and you consistently fall short of the objectives, you will likely be replaced by someone who has proven experience in their ability to make the forecast happen. And if your objective is the same as that of your capital sources, to see the value of the company and your stock increase, you should be glad that you have been replaced. Because truly, there are others who have bonified experience where you may have none. Or perhaps they have a needed skill or perspective that you do not have. Your replacement has done it before and they will likely do it again FOR YOU and your stock value. So being replaced on the management team, or perhaps being demoted is, in the end, in your best interest. Understand this scenario when you get involved with venture capital sources and see such a replacement move as a positive scenario for you.

Raising capital is not an event – It is a way of thinking - You should always be thinking about how you can make your company more attractive to investors. How can I increase the value of my company to provide a better return to my shareholders? If you know your company is a great investment value, you will go into investment discussions with much greater confidence and be able to speak the language of the investor. Want to impress your potential investors? Speak with confidence and speak their language.

You need to be invited to the dance in order to play - Understanding your negotiating position in equity transactions is important. I am thinking of the proverbial entrepreneur who thinks his idea is worth millions of dollars. Ideas are worth zero, nothing. It is only when you put action to the ideas that they begin to develop value.

> **Real Life Experience** – Early on in my career in Silicon Valley, a software engineer and I had developed a new leading-edge security system for financial institutions. We had a prototype and an offer to beta test it in one of America's largest banks. We approached a venture capital firm who eventually presented us with a term sheet. They were of course asking for a majority of the company in return for the seed round of capital. We scoffed at the offer and walked away.

> We didn't realize that we were being offered a ticket to the dance and turned it down. Just the relationship with this firm and the credibility it would have given us as Silicon Valley entrepreneurs would have been worth giving up such a sizable equity percentage up front. So, another start-up that never moved forward due to the arrogance of the management team. Rather than developing a partnership with one of the highest-rated venture capital firms in Silicon Valley at the time, we walked away with our equity, and our pride intact, and no company. As a result of our bad thinking, the company never got started.
>
> The moral of this story is to understand your negotiating position and take the offer to attend the dance. You need to earn your position in the business world. Don't be afraid to take what you perceive as a weak offer if it will get you to the next base in your company's growth.

Honesty is the best policy - I have mentored young companies that do their best to hide their actual performance from their investors. This never ends well. They are your partners and need to be treated as such. If you are having a specific problem, often they will have a solution for you. Look to them for guidance.

> *Real Life Experience* – I was asked by one of the largest venture capital firms in Silicon Valley to work with one of their young start-up companies of five Stanford University software engineers, who had developed a new Java-based web technology called web banner ads, and the VC had taken an interest. They had a good technology but no company infrastructure. I met with the team in their downtown Palo Alto office, and helped them set up their accounting software, operating plan and some basic policies and procedures. I could feel the tension when I would show up at their office,

> but couldn't, at the time, determine a reason for my feeling. They seemed a little anxious to get rid of me. I later found that the VC firm had referred me to them as an independent contractor, yet they felt I was a plant to feed financial progress back to the VC. Such was not the VC's intention nor was it mine. The company was later merged into another investment of the VC firm, as the engineers had a great product, but could not trust the resources they needed to build and sustain their company infrastructure.

A word about giving early equity to participants - As a company with no significant cash resources, you may not have a lot to offer as compensation for labor expended. The temptation is to promise a percentage of the company or to give stock as compensation. My personal feeling is to not offer equity to your early participants. As you will likely see, incoming capital will look very closely at the equity promises you have made in the past as it may represent a significant liability to the company down the road. Investors want to know that the shareholders in the company are all offering some benefit to the company. If early participants who have an equity position, or who were promised an equity position in exchange for labor expended, are no longer contributing to the company, later stage investors will likely not want them in on the deal. These early contributors are receiving a benefit for which they are currently not contributing. So don't cloud your company's equity structure to early.

You may want to look into the *Slicing Pie©* model of early-stage equity compensation. If administered correctly, it may meet a specific need of the company. However, I have no active experience with this equity tool.

> **Real Life Experience** - *I participated in a road show with a young entrepreneur who had a revolutionary process that*

eventually ended up with a first-round funding of $10 million from a notable Silicon Valley VC firm. But before they would invest, they did a review of the company's equity structure and had some caveats for the company president.

The President had promised a significant share of equity to an accountant who had helped develop his financial projections and business plan, early in his company's development. The accountant had been long gone and was no longer involved with the company. The VC firm made the company President go back to this individual and buy them out prior to their providing the $10 million first round. When the accountant was approached, he sensed that there was a significant investment on the table and this left him in a higher negotiating position. Eventually a buyout was structured and the accountant sold his equity back to the company president who bought the equity position personally in order to receive his first financing round from the VC.

Incoming capital has the right to review and change the equity structure to more suit them. Be aware that promises made now will likely have to be revisited and renegotiated at a later point in time.

Real Life Experience – I worked with one company who had informally promised equity, very early in the growth of their company, to a participant. This participant eventually left the company and went on to work elsewhere.

The company went through their formation meeting in preparation for the IPO. The early participant heard about

> the impending financing and wanted his informal promise of equity to be recognized. The company tried to ignore him, but he eventually filed a lawsuit just prior to the IPO. Since all pending lawsuits against the company needed to be disclosed in the IPO prospectus, the management had to deal with this issue to get the lawsuit resolved prior to releasing the IPO. This is the kind of unexpected events you will want to prevent at later stages of development. So even informal promises of equity may need to be resolved later. Be cautious here.

All early-stage equity participants need to understand the concept of dilution before accepting stock as equity. Their starting position in terms of % ownership, will change as additional equity is granted. They need to be willing to accept this dilution as a condition of being granted the stock.

Should I use debt financing? - Taking on too much debt early in your company's development can kill your ability to take on additional capital of any type later on. Savvy venture investors know that the capital they put into a debt laden company will likely go backwards, that is paying off the past debt the company has incurred. When the intent of their investment is to help the company go forward. Their investment immediately gets swallowed up and the company is again in a very high-risk mode, and on its last legs looking for more capital.

If an early investment made was in the form of a loan, accruing interest, the company is incurring even more debt with no way to repay that interest. Large accrued interest on the balance sheet is a definite red flag. Unfortunately, many unaware debt investors will make loans to a company with no plan on how they will get their money, and the accrued interest, back out. In addition, they have helped put the company into a position where they cannot qualify for additional capital.

My experience says that high debt with no, or insufficient business progress forward, indicates an untrustworthy management team. That is why it is important to learn how to move as far forward as you can without the money.

ELEMENTS OF RISK

Why do we care about risk? – In my early years of consulting, I was never aware of the risk an investor sees when considering an investment in a young company. That is until I was responsible for actually writing the investment check. Sitting on the venture capital side of the conference table gave me a totally new perspective on the risk I was taking on, making an investment in a young company.

My objective in this chapter is to let you see your company through the eyes of a potential investor. The investor has to constantly consider the riskiness of the investment and also the potential return on the investment. Remember, the VC or investment bank has a group of investors who have contributed to their fund, expecting a higher-than-average return from their investment. The firm has the pressure of investing the investors' money wisely so they can realize the return expected. If return objectives are not met, the VC firm will lose its business model and its investors.

As a venture capitalist, I had to consider all risks involved and to what degree the entrepreneurial team had mitigated those potential risks. I also had to be able to see a return of our investment and a return on our investment within a reasonable amount of time. So, the pressure was on to find a good productive home for the investment funds.

What follows is a sample of the risks I had to consider in making a potential investment. Read his section with the perspective of your startup and where your risks might be greatest to the investor. And remember it is your duty to mitigate these risks as much as possible prior to an entrance interview. And as a side note, all VC are different and the risks change with the level of investment. So, this was my perspective only, from a seed funding organization perspective.

Old technology risk – Some technologies, though they look good on the surface, may in fact be old technology. Investing in the last generation of technology reduces the useful life and therefore the return on the technology. VC must invest in leading edge and preferably proprietary, technology. To interest a venture capitalist in your company, you must be working on the newest technology available. Remember, technology is like lettuce. It begins to lose its value the longer it is delayed to the marketplace.

Incomplete information risk – One of the reasons it takes so long to get funded by venture capitalists is because they want to reduce their risk by getting to know you and your proposal as well as possible. No investor likes surprises. So, it is imperative that you as an entrepreneur provide complete and verifiable information in your entrance interview and subsequent interviews with them.

Channel risk – Getting your product to market through a distribution channel, means interviewing the first level customer of your products to make sure they are willing to sell and distribute your product to the marketplace. Let me give you an example of channel risk that I encountered early on in my career.

> *Real life experience* - *The entrepreneur had developed a wood-like pressboard product that was composed of straw, an abundant natural resource on the Palouse region of Idaho and Eastern Washington, and HDPE, typically recognized as milk bottle plastic. This plastic is also an abundant resource as it is easily palletized and sold by the*

> pallet. The product was price competitive and as durable as plywood products.
>
> A local lumber yard showed an interest in the product and was willing to stock the product on a trial basis. Later when it came time to ask for a purchase order and deliver the product, the lumber yard backed out of the deal. They said their main lumber supplier would pull their products if they introduced this new product. So, the lumber yard had to decline their agreement to sell the product locally. So competitive issues can quickly clog up a distribution channel that was originally a "no-brainer" as a distribution channel.

Often this kind of risk may not be apparent on the surface. Check your distribution channels and preferably, get at least some purchase orders or at least place some early products into the marketplace to show that this risk does not exist.

Technology leapfrog risk – This risk is similar to the old technology risk. At the time of funding, the technology was the leading-edge technology. However, soon after market introduction, a new, lower cost, likely a more innovative way to accomplish the same end result was introduced to the marketplace, making the initial investment by the investors, obsolete. Other potential competitive technology risks should be reviewed and strategies should be in place if this were to happen. What would be the action by your company to counteract such a competing product?

Pricing risk – Entrepreneurs often price their new product as though everyone agrees with them that they need to have one of these items and they would be willing to pay a higher price to acquire the item. Only to find that the marketplace was not as "in-love" with the product as they expected. So, to sell the product they need to reduce the price. Now the revenue can't be supported that was originally shown by the pro-forma financials. As a result, the financial return that was forecasted will never happen. The investors will likely not be provided a return and the company may even likely implode on itself once its capital runs out.

Disappearing margin risk – Again, the venture investment may have been made on the basis of proforma financials. But inadequate research may have been done up front and the company may find their actual costs to manufacture are higher than forecasted. Now the company faces a reduced profit margin and a lower net income from their product. Again, it is the investors who lose out here as the company cannot provide the returns, and hence the corporate value, shown in their proforma financials. So do what you can to validate your early sales and profit margin.

Alliance risk – One of the quickest ways to get to market is to work with a major market player who has the distribution channel reach into the marketplace. This can provide an immediate market presence for a young company. If such an alliance is made with a small local company, an opportunity may have been passed up to ally with a larger, major market player. Your strategy in choosing an alliance partner should be reviewed in detail. Often an investor can have a relationship with such a major market player and can introduce you to them. Contracting too early with the wrong channel can essentially cost you considerably in your later sales potential.

Team communication risk – In our initial entrance interviews, I would watch very carefully how the team interacted to see if there were any destructive personalities that could potentially cause the blow-up of the company down the road. I would want to see a team that interacted in a very positive, supportive way.

I recall an entrance interview with a team that actually started to argue amongst themselves and did their best to present their point of view by negating the point of view of the other. This was really bad form and needless to say, we passed on funding their company. I would want to see a leader who actually complimented his team. I would look to see that they were organized, and each member of the team showed their particular area of brilliance. A president who did the whole show and never let his team talk, was another red flag for me. It is imperative that the team communicate well amongst themselves and show their flexibility in listening and considering the ideas of others.

Speed/Motivation risk – There is always the risk that the entrepreneur may lose interest in supporting his company, and may decide to move on, leaving the team in a compromised position. So, dedication is an important attribute for a start-up team.

It was also my observation that some individuals and teams could move very quickly to accomplish their goals. Others would take forever to move forward. Start-ups have the ability to move very quickly and this was another element I wanted to see. The team that did a lot of talking but didn't accomplish their goals on a timely basis, in effect, cost the investors a lot of time and money. The ability to overcome hurdles and move quickly toward accomplishing the goals and milestones they set, was a prized attribute.

Divergent/Poor strategy risk – Corporate strategy can make or break a startup. This was an area that often showed the brilliance of the team that I was looking for. Had they arrived at the best strategies for their company in sourcing materials, processes, employees, customers and alliance partners? Did they have a unique angle that gave them a competitive advantage in their marketplace? Competitive edges are critical. Does the company have a well described phased approach to climb the ladder of success, or were they wandering from one opportunity to another? Your strategy is the opportunity to show your brilliance.

Take the money and run risk – Since we were dealing with seed funding, one of my greatest fears was that the entrepreneur would take the money and run to Mexico. Maybe it sounds ridiculous. But were this to happen, I would be the laughing stock of my associates in the other venture capital firms. And would it even be worth hiring an attorney to find the entrepreneur and get the money returned? Likely not. So, I would look for the entrepreneur's stability in the community. Did they have a family locally? Did they own a house? Did they know other people who I knew? Did they go to school locally? Were they involved in social programs with their kids locally? This risk was very real to me and was one reason why so many meetings were to be convened before any funds were transferred to their account. I wanted to know them personally and sense their allegiance to the company and to us as a funding source. You see, my investors were counting on a return from their investment, not a loss of their investment.

Take the technology and run risk – Many lawsuits have been activated because of employees stealing proprietary technology and taking it to another company. Your employees and contributors need to be trustworthy, but you also need to take precautions with non-disclosure agreements. This risk is always present with a young company, as team members can get disillusioned and think they can do a better job with the technology or product and feel justified in stealing it. As a young company with limited funds, you cannot afford to mount a lawsuit should an employee steal your technology. So, take every precaution and monitor your employees' motivations carefully.

Next stage of capital risk – Any investment we made through our VC fund, had to be approved by our investment committee and a larger VC firm, to assess their interest in a later stage of funding. The team would have to show their ability to reach their goals with the seed funds we provided. If the next stage capital was ambivalent and not overly impressed with the investment, we would have to pass on the seed funding. So, getting past the seed stage had to include the potential to interest larger funds once the team had proven themselves with the limited funding we provided.

Dilution risk – If the investment were to "lose its bloom" somewhere along the way, it would become less attractive as an investment. So, later investors would want a more significant position in the company in return for the capital provided. This could happen because of the investment climate changing, the emergence of competitive companies with new products, lack of the team reaching their goals, etc. If the capital was apprehensive about investing, it would cost the current investors dearly in terms of dilution. If the next round of investors couldn't wait to invest, less dilution would likely result and the current investors could remain strong in terms of their equity position.

Liquidity/Exit risk - When investors invest in a company, they need to know how they are getting out. How will they get their investment back and what kind of return will they realize? It is up to the entrepreneurial team to show a plausible plan that will fulfill their investment return goals. Often early-stage companies do not think this far into the deal to consider their investors in this way. So, they take the money and go forward with never a thought about how their investors will achieve a return on their investment. We don't need a rosy picture here about how the market is just waiting for us; nobody has a product like us and we will make lots of money. We need to go to the end and show the investors how they will make a return on, and of, their investment. Appendix B has a forecasting tool you may want to use in this regard.

Product liability risk – Could your product in any way, cause a lawsuit, perhaps through improper use, or accidental situations? A lawsuit presented to an early-stage company will quickly drain their investors' money with no potential for return and can even shutter the company. You need to show your investors that you have considered potential legal issues and done everything to resolve them up front.

Reverse engineering risk – Can your product be easily reverse engineered by a much larger competitor who already has a good distribution channel established? Proprietary elements are highly desired. Patents give a competitive edge and are a marketable item, should the company fold and the investors need to recover as much of their investment as possible. I don't necessarily recommend that start-ups need to have patents to be fundable, but it is one recurring element you will see in companies that get venture funding. If it is easy to reverse engineer or copy a start-ups product, and the company has no proprietary protection, they will need to present a compelling strategy as to why they can achieve their goals in spite of this weakness.

Improper disclosure risk – I hate to say it, but some entrepreneurs are just loose cannons. That is, they can't help but talk all about what they are doing to everyone the come in contact with. Investor agreements are proprietary. They should not be discussed with anyone outside the company unless it is a potential investor well into negotiations. Trade secrets are proprietary. It is important to know what can be disclosed to the outside world and what cannot be shared outside. Improper disclosure can ruin competitive positions, relationships with current investors, and the future of the company.

Employee risk – One of the most important decisions an entrepreneur can make is the decision of which potential employees to hire, and when to hire them. Your employees are the company. Hire the best. Not only in terms of technical competence, but also their ability to flow with the other employees. It is not uncommon to experience employee blow-ups and incur potential legal battles as a result. These are cash absorbers and should be avoided at all costs. Every dollar is valuable to a start-up and must be employed only to help the company move forward.

"Falling in love with the product" risk – Entrepreneurs are often very connected to their inventions. They may even identify personally with the product. But objectivity needs to be maintained here. If someone comes up with an idea to improve the product, it needs to be considered. Even if someone else identified the improvement. Often marketing people can see that some element of the product is not being accepted by the market and needs to be changed. Inventors need to be flexible in product design to meet the needs of the market.

Also, I might add another note in here. VC funding sources will usually provide in their contracts, that a team has to meet certain pre-agreed upon operational or financial milestones. If these milestones are not met for successive periods, the VC reserve the right to put in another company president or technical person to help get the company to achieve its goals. If an entrepreneur is so personally involved with the product, he may have a difficult time letting go to someone else to run "their" company. But bringing in a powerful player who can take the company to a higher position in the industry would be a benefit to them. That is of course if the company president has financial success as their goal and not necessarily being the leader of their company.

If financial success is not your goal as a company, don't partner with someone whose goal is financial success. If the VC determine it is a good time to sell the company, the entrepreneur has to be on board with this decision. So don't take venture capital if your end goal is not aligned with that of the VC.

Arrogance risk – We want to see a leader who is charismatic, and who knows how to enroll and motivate others to accomplish the goals of their company. Unwavering confidence is an admirable and highly desired trait. But no one likes arrogance. It is a pretty quick turn off in the initial interview. If it is something I would pick up in the initial interview, others down the road would likely pick it up as well. And this could impede the success of the company and make working relations with the employees and investors difficult at best.

Improper balance risk –We are all different. We have different skills, perspectives, and frames of reference, and that is what makes a great company, the combining of all of these perspectives and talents.

A talented software engineer needs a good business development manager who has an outside perspective on the company, while the software engineer can focus on the desk in front of him. Without the assistance of a partner with an outward perspective, the engineer's success as a company is at risk.

NEEDS LIST TO HELP YOU GET STARTED

So many entrepreneurs go forward without a plan. They look for opportunities of any kind that will move them ahead. This is not necessarily bad. But to go from opportunity to opportunity, as they happen to arise, can deter you from reaching your well-defined objectives.

We tend to complete tasks we are familiar with and feel comfortable with, while other important tasks may be ignored to the demise of the company. This is one reason we work to a plan. It will help focus us on the path forward rather than taking the path more familiar. I can't recommend this more highly, work to a plan.

> **Real life experience** - *The company president was a natural-born salesman. He was best at talking with potential customers and motivating them to action. This is a prized trait for a start-up entrepreneur. He would send me his weekly status report, which contained information about all of the people he had talked to and all of the potential business he had lined up. But as the President of the company, he also needed to be focused on his capital raise, hiring employees, putting a business development plan together, etc. But instead, he would go wherever he could find an audience. And there were many diversions available when he would work in such a random, undisciplined pattern. His company always struggled from one small dollar investment to another, occasionally missing payrolls, hoping for the next deal to close. I cannot emphasize enough the importance of working to a plan.*

If you are going to go forward in a completely prepared fashion, you will first need to assess where you are in terms of preparation. Below is a "Needs List" that will assess what you have prepared and what tasks you still need to complete. The potential needs of a typical startup are listed in the middle of the list. If your specific needs are not listed, add them to the list.

Observe the first column to the right of the descriptions. This is where you will check off what you need to prepare. If it is an item that does not concern your company, ignore it or draw a line through the description. If it is something you need to prepare, put a check in the "Need" column. Once you have assessed the needs, prioritize them in the second column to the right of the descriptions. As you complete these tasks, you can put the date of task completion in the left most column. Give some serious consideration to this list. You may feel you need it all, or you may feel that you are well prepared and need very few items to complete.

In the pages following the chart, there are the descriptions of the list items that may increase your understanding of the line item and help relieve a little writer's block.

The following "Needs List" may help you prioritize your specific needs and give you a platform with which to launch your progress forward. Often entrepreneurs will stall out with so many needs that they are somewhat shell shocked and need a path forward. Seriously consider using the Needs List below to consolidate and prioritize the current needs. You will be surprised how this can give you a road map forward.

Instructions: Take a first pass of the list putting a check mark in the "Need" column for those elements that you think you currently need to move forward. Then take a second pass and consider which needs are more pressing than the others, by assigning a priority rank in the "Rank" column. As you complete

45

the task to maybe an 80% level, note the completion date in the left-hand column. This will be your progress monitor. As you accomplish the items, you may want to highlight them with a yellow highlighter to show your progress.

Review the list daily to keep your focus on what is important now.

If you are seeking seed funding, discussing this chart, and the results achieved, with a prospective investor will gain you credibility in that you are identifying and prioritizing needs and you are showing progress to acquire resolutions to these needs. Any time you can show a plan and your successful progress against that plan, it will be a significant upgrade in the eyes of an investor.

NEEDS LIST

Complete Date	Description	Need	Rank
		✓	1,2,3
	Elevator pitch		
	Concept paper		
	Company exclusive bank account		
	Company Credit/Debit card		
	Strategic plan		
	Operating/Financial plan		
	Product generation/expansion plan		
	Marketing plan		
	Business Plan		
	Marketing specialist (advice)		
	Marketing partner		
	Development partner		
	Technical partner		
	Equipment		
	Raw Material		
	Office space		
	Lab space		
	Manufacturing/production space		
	Retail space		
	Test equipment		
	Referrals		
	Software/Hardware Engineer		
	Scientific assistance		
	Web help		
	Prototype		
	Accounting system		
	Accountant		
	Purchasing assistance		
	Licenses – Business, etc.		

	Mentor		
	Partner		
	Advice		
	Advisory board members		
	Sales presentation material		
	Intellectual Property		

Elevator pitch – Give some thought as to how you would describe your company and product. A one sentence description and then perhaps a one paragraph description that you can memorize and recite when asked.

Concept paper (sample in appendix A) – It may be helpful when approaching potential contributors, to have a one or two-page summary of what you are trying to accomplish and perhaps a short list of the resources, personnel, or tangible resources you are looking for. This gives a potential contributor a "way to play" with the team. If they see that you are looking for a bookkeeper and they are a bookkeeper, they can immediately visualize themselves as having a way to participate in the company with you. And again, if you are asking for maybe a couple of hours of help per month, you would be surprised at the response you may receive.

Also, whenever you talk with someone about your endeavor, leaving behind a printed piece that explains the company concept and the resource needs you are looking for, provides a presentation piece that can work for you even though you are not present. The concept paper can be handed to another individual who may show an interest. This also shows you are prepared for the discussion. No, "I'll send you an email". Here is the presentation, now is the time. Emails are easily overlooked or passed by if the interest from the discussion is not maintained.

Company exclusive bank account – Part of the IRS (and GAAP) requirement for your business is that you maintain a bank account exclusively for your business. No personal purchases or withdrawals from this account, exclusively business transactions. Setting this up sooner rather than later will make the accounting much easier and will keep all of the bookkeepers needs focused in one place. A point of information here, you typically cannot open a bank account without a business license, and you can't get a business license until you have registered your business name with the state business name registry. So doing this first will save you another trip to the bank later.

Another note, do not make any cash withdrawals from this bank account. A good set of books shows an audit trail back to each business expenditure made as evidenced by an invoice or receipt. When cash is withdrawn, this audit trail is lost. This introduces an element of doubt into your business transactions, not to mention making it difficult for your bookkeeper to know how to post the withdrawal.

Company exclusive credit or debit card – Again, comingling personal and business credit card purchases produces a nightmare for a bookkeeper and may invalidate your corporate formation in a court of law. Keep a credit or debit card exclusively for business purchases.

I have been called in to help do the accounting work to extricate personal expenses from business expenses. Yes, I can do this, but it rarely turns out good. You will have a significant weakness in your financials going forward if this needs to be done at a later date.

Business plan – Outside stakeholders want to know that you have a good perspective on your business and have a clear path to profitability. A good business plan is important, as your business ability will be judged on how well you have presented your business in the plan. It is also a document that must speak for itself when you are not present to answer specific questions. I have read hundreds of plans and will offer some advice that may be helpful for you.

First off, please do not use a business plan template. If you are having writers block and don't know what to say, you may look at some templates to give you some ideas. But please do not use a template to prepare your plan. I could immediately spot the plans that used a template, and most of them would get put to the back of the line. I would look for certain weaknesses and if I saw them, I would probably pass on the investment.

Every business is different. This difference requires that the plan be customized to answer the most prescient questions generated. For instance, if our marketing plan talks only about your competitors and why your product is better, you have missed the whole point of the marketing section.

Address the issues specifically related to your company and its products. As you make public presentations of your company, be aware of the questions asked and make sure you address these questions in your plan.

I would always favor the plans that showed some strategic depth. Don't just tell me the prices of your products. How did you arrive at your pricing structure? What later generations of product will you design and produce? Have you considered the potential effect of your competitors taking a retaliatory response to you entering the marketplace? What happens if your competitor starts to also produce a product similar to yours? Want to give your investor a good feeling about your plan? Let them know you have thought through these deeper issues and have answers for their questions.

You might consider conducting a role-playing session with someone who has a business mind, someone who could uncover questions that might be asked that you would have never thought of previously.

Strategic plan – Your strategy should be revealed in your business plan. This will of course make your business plan somewhat confidential. So only reveal the business plan when it is necessary to do so if it contains confidential strategy or information.

Operating/Financial plan – Being a VC finance guy, I would go immediately to the proforma financials in the business plan. This would tell me a lot about the true ability of the entrepreneurial team. Does your balance sheet balance? Have you captured all of the typical operating expenses I would expect you to incur? Are your cost of goods sold reasonable according to industry averages. If not, why are they credible though different? Do you show an inordinately high revenue your first year? If so, can your production and business development strategies support such a revenue?

By the way, 95% of the business plan financials I read, significantly overstated early-stage revenue. It takes time to get your company and its products noticed by the marketplace. So, your promotional strategy has to correspond to your revenue projection. Realize that when you put out a revenue projection, you are establishing a performance milestone. If you miss the revenue milestone, your financial reporting integrity will be lost.

Remember that often a venture capital contract will state that you must meet the revenue or income metrics shown in your plan. If you do not perform to plan, you will be at risk of being replaced by a more senior professional who the VC believe can make the forecast. So, show whatever revenue you truly believe you can achieve. But be aware that not meeting your proforma projections can cost you dearly in terms of your position with the company or your equity at a later stage. Be conservative here.

A good operating plan will be resource-based. That is, it will show what personnel positions you intend to fill and when you intend to fill them, what expenses you will incur and when you will incur them, what capital equipment you will need to buy and when you will need to make these purchases, etc. In short, a good financial operating plan can answer a lot of questions relating to what, when, where, why and how, saving you the need to mention such detail in your business plan. The operating plan also pulls all elements of your future plan into a financial picture and includes the timing of all operations. Without such a plan, you go into the future with no plan, just a list of hopes.

I would often find it interesting to ask an entrepreneur how much money they need and what will they do with the money, what milestones will be achieved? The answer was typically expressed in generalities, since they know they need more money, but are not sure of how much. Again, a red flag for me. Tell me what you need, how you are going to spend it and what milestones you expect to reach with the funding.

Appendix C shows the typical content of a good operating plan. Actual Microsoft Excel models of two good operating plans can be downloaded with the hotlink in Appendix C. It would be helpful to have some accounting knowledge if you engage the full operating plan model. The short model is more appropriate for a user with limited accounting knowledge.

Product/Generation expansion plan – Equity capital invests in companies, not initial products. So, you should have a plan for the extension of your technologies into later products. This needs to show the building of a company in terms of product milestones. Yes, you have a great product today, but what additional improvements or products will you release down the road to continue your corporate growth. Being focused on one product leaves you at risk of not being able to present your company beyond its initial products or services. Determine if a product generational plan would be useful for your company.

Marketing plan – Address the 4 P's and you will have covered most of the marketing perspective of your company. Generally, these are Product, Price, Promotion and Place. Again, some companies depend very heavily in a successful marketing plan, others, not so much. If it is important to your business planning, then you need to address this issue. One size does not fit all here in terms of a standard business plan.

It will be helpful to consider an early-stage roll-out plan. Once you begin selling to your customers, how will you get them to respond? How will you get noticed and how will you succeed in getting them to buy your product or service? This is key to being able to defend your early-stage revenue forecast and your marketing budget.

Marketing partner – Give some consideration as to whether there is another company with whom you could partner, to bring your offering to the marketplace. This is a strategic decision involving many issues specific to the competitive environment, costs of market introduction and depth of reach needed into the marketplace. Sometimes it is best to "go it alone" and other times partnering is a great way to gain an instant market presence.

Development partner – Likewise, if your first-generation product will require development into other stages/products, it might be helpful to partner with a firm or individual that has the technical experience and perhaps the capital equipment to help develop your later generation products.

Purchasing high-cost equipment – This is a difficult category that often causes the funding requirements be so high for a startup. It also represents an element of risk for investors. Consider if there is another way to acquire the result the capital equipment would provide, without paying for the equipment or finding a creative way to finance the equipment. Who else has that piece of equipment you need and is there a way you could use the equipment in the early stages of your company for prototypes, initial product runs, or perhaps even an ongoing production relationship?

If you came to me looking for equity capital, and I saw that you had reduced the need for investment capital by finding a creative way to use someone else's equipment and eliminate my money going to a high-cost piece of equipment, I would consider that as a significant strategic move, gaining you a significantly better chance of receiving the initial capital you are looking for.

Raw material - Acquiring raw material may be a challenge for your company, particularly if you are sourcing a rare commodity. If this could be an issue, it would be comforting to know you are aware of the issue and have a strategy to guarantee a continued source of the commodity. If it is an area that needs specific focus, be sure to address it in your plan.

Office/Lab/Production space – Your startup may require specific space from which to work. If you are a retail operation, you may need a storefront. If you manufacture a product, a warehouse off the beaten path may be appropriate.

Many businesses have facilities that appear to be fully utilized, but are often not. Inquiring as to subleasing or renting a small space on a month-to-month basis within a larger building may be a cost-effective way to get started. Rental expense and lease contracts are a significant cost to the startup. If you can start virtually, do so. If you need the space, use some practical sense here.

I would highly recommend, that if you are not yet generating revenue, don't commit to a lease on a facility that you "could eventually grow into". I have watched many Silicon Valley startups who got their first round of funding, tie into 5-year leases into top-floor quality office suites when they haven't even proven that the market wants their product. This often leads to bankruptcy. Don't commit to lease contracts unless you are sure you can fulfill them. And don't go high-profile just to attract capital. Often it is a signal of your bad judgement to be overcommitted to space that is occupied to impress others.

Again, if you are a student, often universities provide space for budding enterprises. The resources in a university setting are abundant. Use them if they are available and you have the need for the space.

Test equipment – Often this equipment can be rented or even borrowed. Again, if anyone in your area has the test equipment you need, see first if you can borrow it to run the tests you need. As another option, electronic test equipment can be rented on a month-to-month basis, preserving capital.

Referrals – Referrals are one of the most valuable tools for a start-up. If you need to get in front of those who can help you further your endeavor, a simple referral could make or break your start-up efforts. Be prepared to ask those with whom you meet if they know of someone else who could help move your company forward. The worst that could happen is they say no. The best that could happen is for you to find an avenue "out of the darkness" and into the light.

Software/Hardware engineering – If you need a partner with specific expertise, note this in your needs list and begin looking for that individual. Consider if you can buy software code or if it will have to be fully developed from scratch. Having a block diagram, no matter how unsophisticated, which explains specifically the software tasks that need to be accomplished, can be helpful when talking to technical experts and funding sources.

Web help – A generic web site is fairly easy to develop by anyone. But if you require a database integration, live feeds and other such options, there is a technical expert out there for which this is very easy to do and can be done relatively quickly. Think ahead as to specifically what you need for this generation of product, as well as successive generations. Detail out what your company and products will look like several years from now so the developer can build the initial stages of your website with these generational enhancements in mind.

Prototype – If your product can be displayed as a prototype, consider early how you will get this prototype built. How can your bench proof-of-concept be shown in its first-generation product? Use some creativity here as you may not necessarily need a final production process to produce a look-alike prototype. Taking your technology from the bench to a prototype is a major step that will help all potential stakeholders envision, and relate to, the product.

Accounting system/Accountant – Being an accounting and finance guy, I have been called into many startups after they have been running for a while. The management hasn't understood the importance of good accounting records and have let this discipline fall by the wayside. I understand employing accounting help can be an overhead expense burden that is not required to move the product forward. But remember, investors invest in companies not products. Hiring even a low-cost accounting resource on a part time basis will be well worth it when it comes to arranging funding down the road.

Partner – One plus one equals three. Look for a partner who has a passion for the product similar to yours. There are many benefits to partnering and having another point of view. Often partnering with a student will allow access to university resources.

Advice – Information reduces uncertainty. Always be open to advice from an expert or even another concerned party. Determine exactly what advice you need and seek out the individual(s) who can help you.

Advisory Board member – You may want to give some thought to developing an advisory board. This is not a formal board of directors, just a list of contributors who are willing to meet with you periodically to review progress or to offer specific assistance where needed. Also, give some thought as to how you intend to compensate them with an occasional reward of some type. A bottle of wine, concert tickets, dinner, an offer to help them with a need of theirs. Even a note of appreciation is welcomed by such contributors. I would recommend a one-year term for members. Those that are helpful can be extended, but often some advisors become disinterested or do not want to be listed as a contributor. So, adjust here as you see the need.

Sales Presentation Material – When you eventually get into the presence of a potential client or customer, what documentation will you need? Plans, pictures, samples, drawings, copies of articles, contracts? Any presentation made, needs to have leave-behind documentation, as the person who received the presentation will not always be the decision maker. Yet the decision maker needs to have the information to allow them to evaluate and make a final decision. You can save yourself recurring visits to explain your product and its benefits if you have the next steps already presented, leaving the next action in their hands. Think ahead into this process and you will accelerate the close and decrease your sales cycle.

Intellectual property – IP includes patents, trademarks and service marks. Carefully consider your need for these registrations. Realize that a patent is only as valuable as the money you have to defend it. Legal costs to get the patent, not to mention the cost to defend it, can be considerable and are often beyond the reach of a seed funded company.

A DISCUSSION ON CORPORATE VALUE

A well-planned start-up will have set some goals and then set some milestones along the way to check progress to those goals.

You need a corporate plan - I cannot emphasize how important a corporate plan is to guiding your company towards its goals. Oftentimes I will see young companies focusing on getting to market with their product(s). No question that revenue is the prime driver of corporate value. So, there is not necessarily any error here. However, your goals should be much broader in terms of your corporate development.

Often a company without goals is simply looking everywhere for opportunity and will look at success as having established contact with a new provider of value. When as valuable as this new provider may be, focusing there is a diversion from accomplishing the goals necessary to provide maximum corporate value.

Your ultimate goal should not be revenue. Your ultimate goal should be establishing a return for those who have invested in your company. Certainly, this cannot be achieved without revenue. So, revenue is not a bad place to start. But it is a bad place to end in your planning.

If you can step back from your company and look at it as an entity independent of you, then maybe you can more accurately envision what the company needs to move forward. You will build the value of your company by enrolling the right resources, at the right time, to increase corporate value.

The best overall corporate goal I have seen, is to provide a return on, and return of, the investors capital. This should be the first priority of any company that has accepted an outside investment, whether it be a loan or an equity contribution. Often this goal will be realized through successful operations, but perhaps also through a sale of the company. You should think through the way you will achieve this goal, and you should be able to explain it to your investors, before ever receiving an investment in your company.

Stock value and the achievement of milestones – I have always related corporate value (stock value) to the management team accomplishing specific milestones. Only once the milestones for this stage have been realized, is the company qualified to go out to raise another round of capital, that is to sell additional stock at a higher price than the previous round.

Further rounds of capital are not guaranteed just because you need the money. You need to justify why the new round of capital raise will be sold at a higher price than the last round.

I would illustrate this with the attached worksheet as Appendix B in this book. (The numbers and milestones shown are only placeholders. The real metrics would need to be specifically identified and agreed upon by the management team at the start of the project). A Microsoft Excel version of this worksheet is available as a free download using the links provided in Appendix B.

The worksheet shows a projection of the different rounds of capital to be raised, the price for the stock sold and the total capital raised. Notice the column on the far right. These are examples of milestones that need to be reached by the management team before they are qualified to sell the next round of stock. This gives the team an understanding of the relationship between stock value and milestones accomplished.

This is an example of meeting the needs of the investors, showing them how they will recognize an increase in the value of their stock over time. Never once in my career, did I find a young company that had thought this far into their capital plan. Give it some consideration and you will be the company that attracts the investment capital. Then it will be up to you to accomplish the milestones with the capital raised.

CONCLUSION

Successfully starting your company is really all about a positive state of mind and about establishing a clear perspective and generally a healthy way of thinking from the beginning.

You need a corporate plan to guide you to specific goals so that you do not waste time on pseudo-goals that "look promising" but will not move you forward in the direction you need to go to optimize your corporate value.

Hopefully the insights above will help you to maybe rethink the path forward.

I would welcome hearing your comments and hearing of your similar experiences.

Appendix A
Sample Company Concept Paper (fictitious)

ElectronTransport, Inc.

Transporting human life through cyberspace

For additional information please contact:

Ron Johnson
1212 N. Washington Street, Suite 130
Spokane, WA 99201
(916)524-3999

Rev 1.0
For Information Only

This memorandum is neither an offer nor a solicitation of an offer of any investment in Electron- Transport, Inc. An offering can only be made pursuant to an offering circular in accordance with federal and state securities laws. It is for confidential use only and may not be reproduced, sold or otherwise redistributed without prior written approval of Electron Transport, Inc. Any action contrary to these restrictions may place you and the issuer in direct violation of federal and state securities laws.

This concept paper has been created for educational purposes only. Electron Transport, Inc. does not exist nor do the products or technologies presented.

©1999 Electron Transport, Inc., All Rights Reserved.

The Situation
The technology is now available to transport human life through cyberspace. The technology has actually been available for several years through independent stand-alone products. It is through skillfully combining these products, developing the proper interfaces, software code and transport equipment that electron transport through cyberspace can in fact become a reality.

The technologies that make this project possible include the common laser, holographic imaging techniques, microwave communication and a Magnetic Resonance Imaging (MRI) scanning platform. Though the conceptual proof of the product will require a uniquely and specifically designed transporter room, we expect the technology to eventually be developed into a PC computer add-on.

Electron Transport, Inc. will develop the transport construct over a 12-month development period and expects to prove the feasibility of the science through a prototype demonstration at the end of this 12-month development period.

We have contacted Stanford Research Institute and they have agreed to become a development partner to Electron Transport, Inc. Dr. Werner Von Braun has also agreed to act as a technical advisor to the project.

The Need - Management
Prior to raising the money required to fund this project, we need to build a management team here in Spokane, Washington. Specifically, we need:

- Director of Software Development – To coordinate the writing of the front and back-end systems and the "glue

code" which will be required to integrate the independent technologies.
- Director of Human Resources – We expect to ramp the company headcount rather rapidly and will need a Director of HR to coordinate the hiring and maintain the HR function.
- Chief Financial Officer – A seasoned financial type is needed to prepare the business plan and specifically the financial statements which can interpret the strategies of the company.
- Nuclear Physicist – This key position will ensure that the transport mechanism is 100% safe and that we will not lose any of our clients in cyberspace.

The initial management team will not be required to commit full time until the funding is in place. Salaries will not be paid until the funding is in place. Founders stock will also be issued to the team at this time.

The Need – Advisory Board Members
In addition to the management positions, we are seeking key advisors as board members of the Technical Advisory Board. These board members will contribute through a monthly meeting where their advice will be sought. Advisory Board members will not be paid, but will receive Electron Transport stock for their contributions.

If you are interested in initially participating in this start-up, please contact Ron Johnson at the phone number or email given on the front of this concept paper.

First Phase - the transporter room

The first phase of the project will be the development of a transporter room in which the technology will be developed and proven. This room will be housed in its own building due to the high-energy microwave power generated and used in the transport system. Due to seismic requirements, the transporter building will be located underground, somewhere in the Nevada desert. Specific location will be held confidential due to the political implications of the development of this project. Initial transports will be from ground zero, current time and location, out into cyberspace, and back to ground zero.

Appendix B
Equity Growth Chart
Blue numbers are user input. A MS Excel version of this worksheet can be downloaded from
https://www.theinterimcontroller.com/equity-plan

Your Company
Equity/Milestone Worksheet — For Illustrative Purposes Only

Draft 1.0

Stage / Stockholder	Type	Seed Shares	Seed % (Post)	A Round Shares	Rnd 1 % (Post)	B Round Shares	Rnd 2 % (Post)	C Round Shares	IPO % (Post)	Amount Raised $	Milestones:
Seed											
Founder	Common	2,000,000	33.3%	2,000,000	22.2%	2,000,000	17.4%	2,000,000	9.3%		R&D Complete
President	Common	800,000	13.3%	800,000	8.9%	800,000	7.0%	800,000	3.7%		Patents Filed
VP Marketing	Common	800,000	13.3%	800,000	8.9%	800,000	7.0%	800,000	3.7%		Partnership agreements signed
COO	Common	800,000	13.3%	800,000	8.9%	800,000	7.0%	800,000	3.7%		Successful Beta site
CFO	Common	800,000	13.3%	800,000	8.9%	800,000	7.0%	800,000	3.7%		First 10 product sales
CTO	Common	800,000	13.3%	800,000	8.9%	800,000	7.0%	800,000	3.7%		Business and Operating Plan complete
	Common		0.0%		0.0%		0.0%		0.0%		Successful product launch
	Common		0.0%		0.0%		0.0%		0.0%		Pro-forma BOD formed
Employee Stock Options	Common		0.0%		0.0%		0.0%		0.0%		Rnd A investors identified
Total Seed Funding		6,000,000	100.0%	6,000,000	67%	6,000,000	52%	6,000,000	28%	$ -	
Price per share		$ 0.001									
Valuation		60,000									
A Round				Shares							
A Round Investors	PfdA			3,000,000	33%	3,000,000	26%	3,000,000	14%	1,500,000	$1 million in revenue
	PfdA				0%		0%		0%		42% gross margin
	PfdA				0%		0%		0%		Production floor expansion complete
	PfdA				0%		0%		0%		Attend 3 trade shows
	PfdA				0%		0%		0%		Hire VP of Engineering
	PfdA				0%		0%		0%		Rnd B investors identified
Total Round 1				3,000,000	33%	3,000,000	26%	3,000,000	14%		
Price per share				$ 0.500							
Valuation				4,500,000							
B Round						Shares					
B Round Investors	PfdB					2,500,000	22%	2,500,000	12%	5,000,000	$10 million in Revenue
	PfdB						0%		0%		50% gross margins
	PfdB						0%		0%		European expansion in place
	PfdB						0%		0%		Product upgrade 1 complete
	PfdB						0%		0%		3 new products introduced
Total Round 2						2,500,000	22%	2,500,000	12%		New facility under contract
Price per share						$ 2.000					Partner for market segment 3 under contract
Valuation						23,000,000					Rnd C investors identified
C Round											
C Round Investors	PfdC							10,000,000	47%	$100,000,000	$100 Million in sales
	PfdC										50% margins
Total Round 3								10,000,000			2 acquisitions complete
Price per share								$ 10.000			Market domination
Valuation								215,000,000	Exit Valuation		
Total Shares Issued		6,000,000		9,000,000		11,500,000		21,500,000			

67

Appendix C
Operating/Financial Plan Overview
A Microsoft Excel version of this model can be downloaded from:
https://www.theinterimcontroller.com/operating-plans
or:
https://sites.google.com/view/excelmodelsforpssclnmbook/home

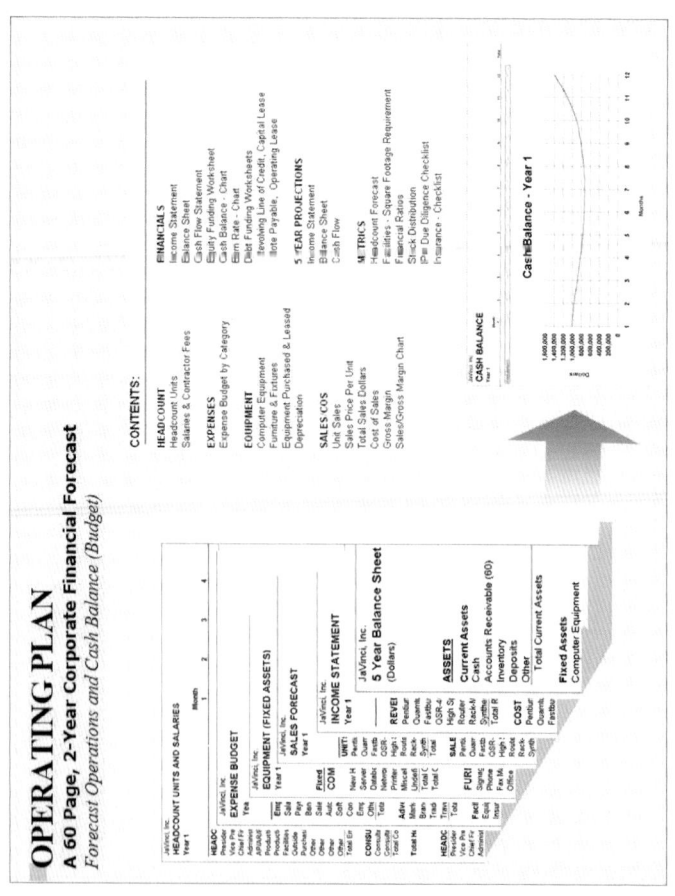

Appendix D
Operating/Financial Plan Resources

A Microsoft Excel version of several free corporate planning models can be downloaded from the Interim Controller website as below. These planning models do not use any advanced Excel functions, straight simple Excel. And as such, they can be easily modified to fit your particular company.

Full Operating Plan – This plan requires a bit of accounting knowledge to complete, but provides a wholistic approach to corporate planning. All resource changes to the model back into your bank cash balance. So, the objective of the model is to prove your business model and justify how much financing you need and how it will be spent. I have used this model in over $1.2 billion in startup financing while working in Silicon Valley.

Short Operating Plan – This plan does not require any specific accounting knowledge.

Downloads:
https://www.theinterimcontroller.com/operating-plans
or:
https://sites.google.com/view/excelmodelsforpssclnmbook/home